PREFACE

Allis Chalmers as a tractor manufacturer in its own right is but a memory. This is the we have produced on the most popular models built by that concern.

One important point to notice is when comparing production figures of A-C tractors these seem very small compared with the giants e.g., John Deere and International, but it has to be remembered that A-C activity took them into many fields of engineering and construction.

This title has been reworked and additions made to make it in concord with other recent A4 titles on American makes.

Finally, we would point out to readers that the author of this book aims to give a fair and unbiassed account of any manufacturer's models unaffected by any fettish or favour towards any particular make!

ACKNOWLEDGEMENTS

Bill Huxley, Bob Moorhouse and Charles Cawood are to be thanked for their assistance in supplying material to enable this book to be put together.

ISBN 0 907742 61-0

Design, layout and editing: Allan T. Condie

© Allan T. Condie Publications 1988. All rights reserved.

No part of this title may be reproduced in any form by any means, including photocopying, except for normal review purposes without the written consent of the Publisher. It shall not be disposed of at any other cover price than that shown thereon, or in any cover or binding other than that it was first supplied.

Allan T. Condie Publications, "Merrivale", Main Street, Carlton, NUNEATON CV13 0BZ.

This picture says it all! A Model 'U' dating from 1935 in charge of a "Cutlift" harvesting clover. Note the hitch to enable the trailer to be hauled by the same tractor.

The Allis Chalmers Tractor Story

"U" – "United"

Some 32 indepentant manufacturers and distributors of tractors, farm machinery and industrial equipment combined in a co-operative manufacturing and marketing organisation in 1929 with headquarters in Chicago. A complete range of farm equipment was to be offered, to be used in conjunction with the co-operative's own tractor, the "United". This was built under contract by Allis Chalmers, who had joined the co-operative. It appeared at the Southwest Tractor and Road Show in Wichita, Kansas, in March 1929.

Constructed on the unit principle, it owed its ancestry to the earlier models produced at Milwaukee, but incorporated a four speed transmission and used a side valve engine of continental manufacture with 4 cylinders of 4¼" bore by 5" stroke developing 30HP.

The inevitable happened, and the co-operative folded, due mainly to petty disagreements between various of its constituents. Allis Chalmers adopted the United however and with merely a change in radiator top tank and side panels it became the Allis "U".

Nearly 2,000 "United and U" tractors were produced in 1929 and the total number produced with Continental engines was 7,404. In 1932 Allis Chalmers produced its own UM engine of overhead valve design with 4.375" bore and 5" stroke, rated at 34HP and equipped the U, UC and M models with it. It was one of the most successful spark ignition engines ever put into a tractor, and unquestionably the best governing tractor engine of all time!

In 1936 the "U" was updated and the driving position moved to a more rearward position. At the same time the engine bore was increased to 4½" giving a rated belt horsepower of 39.5.

The "U" had several other claims to fame however. Allis Chalmers in conjunction with B.F. Goodrich & Co. (Goodyear) experimented with a model "U" fitted with low pressure aircraft tyres, subsequently devising the familiar lugs seen today in some shape on all tractor tyres. A-C offered pneumatics as standard on their "U" late in 1932, and to publicise the fact organised a publicity campaign which was to make the rather ordinary "U" famous.

One of the obvious advantages of pneumatic tyres over steel wheels was the fact that the tractor could be driven along roads without damaging the surface. A-C played on this by fitting model "U" tractors with special rear axle parts and used these at state shows and State Fairs for racing, employing well-known racing drivers to show their prowess in handling "U"s at speed.

Barney Oldfield was one of these, and he drove a "hotted-up" "U" over a mile course at Dallas, Texas, at an average speed of 64.28mph, thus establishing a world speed record for tractors. Later, Ab. Jenkins bettered this by driving a "U" at an average speed of 65.44mph at Firestone's Farm, Colombiana, Ohio, a record which, as far as is known, has not yet been bettered.

All these goings on did much to convince the farmer that pneumatics were the thing to have on his tractor. Although the standard "U" could only do 12mph in top the UI industrial could reach 30mph.

Allis Chalmers did not manufacture its own wheels or wheel centres. The well known supplier French & Hecht provided these exclusively until the late '30s.

The "U" had been imported into the UK in small numbers from the mid '30s. With the "lease-lend" arrangements during the Second World War a number of "U" tractors were imported and did much to endear that model to British farmers. When Allis established their Totton works after the war the "U" was offered until 1950. Whereas during wartime, tractors were shipped whole; the latter day output from Totton had been assembled there from crated form.

The best remembered features of the "U" are the glass jar "Ball Special" pre-cleaner on the air intake stack, the ineffectiveness of the primitive band type brake when wet, and the almost silent tickover which erupted into a loud bark when the throttle was opened. Although some "Us" were fitted with adriolic lifts in the early '50s the tractor was never offered with any type of lift either in the USA or Britain as a production option. In all 22,000 odd "Us" were produced.

"All Crop" and "UM"

Allis Chalmers entered the "General Purpose" tractor market in 1930 with the "All Crop". This used the same continental engine as the "U" and some common components but the rear axle transmission was of different layout. Like the "U" it was advertised as a "Three Plow" tractor but it was also expounded that it "could be hitched or unhitched to or from its cultivator in five minutes."

The "All Crop" received the UM engine in 1933 and became the UC. It underwent a certain amount of restyling in 1936 at the same time as the "U". Pneumatic tyres were offered from 1933 and production continued until 1941 when the model was dropped apart from the supply of certain skid units to the Thomson Coy. for use in "Cane" models.

For some reason the "UC" never sold in the quantities of its rivals such as the Case "CC", Oliver 80 Rowcrop, or International Farmall 30. Some 5,000 were built in the 11 years of its production, along with a whole range of mounted implements to suit, the tractor being equipped with a power lift. No UC tractors are known to have been imported into the United Kingdom.

The Thomson Machinery Co., of Labadieville, Louisiana produced all types of equipment for the Cane fields. Although they converted other makes of tractor the "UC" was most used as a basis for a Cane tractor well into the '50s. Even after the "UC" production had ceased Thomson took skid units, finally adopting the GM2-71 diesel as a power unit.

The 4-wheel drive adaptation used a "U" transmission at the front and a "UC" at the rear. High clearance was necessary for Cane work, and a heavy duty hydraulic lift frame was normally fitted at the rear. Thomson moved away from A-C in the early '60s and tended to use more Fordson skid units.

The basic "UC" Cane received 9.00 x 36" rear wheels and a wide arched front axle to give the necessary clearance. Early models used the mechanical lift (adapted) but engine driven hydraulic pumps and dual ram type lifts came in the early '50s.

"The Popular "WC"

1933 saw the introduction of one of the most successful models to come out of West Allis works at Milwaukee. The WC "Two Plow" General Purpose tractor. The "WC" had the distinction of being the first General Purpose tractor on rubbers to be tested at Nebraska. The tests showed a HP at the drawbar of 24, from a 4-cylinder engine of 4" bore by 4" stroke. A power lift could be fitted as an option.

The WC was the first of the line to be styled in 1937. Some 175,000 "WC" tractors were built, the model being updated in 1948 (c.f.)

The illustrations overleaf show the variety of wheel equipment which was available with the WC. Numerous attachments available for mounting on the tractor included tworow planters, cultivators, rotary hoe and mulcher.

All of the implements for the tractor were designed so that it could be driven into them and hitched. Many General Purpose models required a considerable amount of work to fit mounted implements – not so the Allis range – one just drove in!

Whereas the "U" used unit construction a sub frame was adopted for the "WC" and this enabled the gearbox and final drive assemblies to be as compact as possible, their design following more of automotive practice than that of tractors.

The "WF"

The "WF" was the 'standard' version of the "WC" and was produced from 1937 to 1951. A "WI" industrial version was also available. The "WF" was not styled until the very end of 1938. The engine, gearbox and transmission were the same as the "WC" but the final drive casings were of different design to bring the height of the tractor down. Although "WC" and "WF" tractors were common in the UK, the later "WD" model appeared only in the Irish Free State.

The "WF" was rendered superfluous by the movement afoot in the late '40s which combined the design features of "General Purpose" and "Four Wheeled" units. With the "WD" available as a four wheeler there was no real need to continue a separate variant now that mounted implements were coming into vogue. The WD45 adopted three point linkage style of hydraulics which were becoming universal in the mid 1950s.

The Outdated "E"

The Model "E" owed its origins back to the original 1919 18-30 model. It used Allis Chalmers' own engine from the outset when it appeared as the 25/40 in 1930. It was handicapped however, by the use of an outdated transmission layout compared to the "U" and, the excellent engine, which was also adapted for the "35" crawler in the same year, provided 40HP at the belt. The standard engine was 5" bore by 6½" stroke but as the same engine with bore enlarged to 5¼" was adopted for the K crawler it became optional with the model "E" – "this larger engine is furnished only with tractors that are to be used exclusively for threshing or other work" – was stated in contemporary literature.

Two forward and one reverse speeds were provided. From 1932 the tractor was offered on "air tyres" although the slogan above the contempory sales literature "no more keep of the pavement signs" was hardly apt for a machine which was unable to exceed 3½"mph.

Some 1,500 model "E" tractors were built between the years 1930-36.

Compromise: the "A"

To fill a gap when the model "E" was dropped in 1936 a tractor of similar power output an "A" was introduced. It used the same engine as the "E" but with bore reduced to 4¾", although a 5" bore engine was available for distillate operation. The "U" rear axle and transmission was taken, beefed up and mated by means of an adapter flange to the bigger engine and the rest of the tractor was a curious adaption of "E" and "U" styles.

The resultant beast weighed some 3¾ tons in working order. Some 1,200 units were built and the model dropped in 1941.

A number of model "A" tractors were imported into the UK during 1941 and these have the distinction of being the most powerful model wheeled-tractor to be imported during the Second World War. 50.5HP was available at the belt and on test the figure showed 62BHP, compared with the 45 of the Case LA and the 53 of the M-M GT.

Although dubbed a "Four Plow" model, the "A" gained only limited popularity mainly with threshermen, and had a reputation for being a fuel eater. The slightly smaller "U" would do almost the same work and with half the fuel.

The Ubiquitous "B"

The most familiar model to UK farmers was, of course, the "B". It originated in 1937 as a light general purpose machine featuring unit construction by adopting a reinforced "torque tube" to join engine and gearbox. This gave the tractor a somewhat 'wasp'-like appearance.

Initial production in 1937 used a Waukesha type FCL engine of 3" by 4" stroke, this accounting for the first 96 units. Allis Chalmers then fitted their own engine of 3¼" bore by 3½" stroke which was used until 1944 when the bore was increased to that of its allied 'three wheel' model the "C" to 3.375".

A number of "Bs" were imported into the UK between 1938 and 1941. The tractor was available with bowed and adjustable straight axle styles from the outset, the latter type being used on two derivatives the "Potato Special" and the "Asparagus Special". By fitting larger rear wheels a higher clearance was available for asparagus cultivation. The "B" was available only on pneumatic tyres. As with all Allis Chalmers models, the "B" was available with self starting and lighting from the outset.

Despite its size, the B developed 16HP at the belt and could handle "One Plow".

An industrial version of the "B" was produced and no airport in the USA at one time was without one. The "B" lasted longer, unmutilated in the USA, and by 1957 when production ceased nearly a quarter of a million units had been produced. Later models featured improved brakes, coil ignition, and a hydraulic lift for implement control.

The "B" had not been available in the UK from 1942 and it became somewhat of a surprise to learn that the "B" was to be assembled at Totton. Totton was a 500-acre site near Southampton acquired in the '30s to develop fully the UK side of the business. In practice the British "B" started off as little different from the American one, and in fact most components were imported initially. As time progressed the purchase of units from UK sub-contractors increased however.

In the early '50s various manufacturers had conversion kits on the market to enable the rockshaft type lift on the "B" to be adapted as three point linkage yet it took some time before Allis Chalmers produced this as an option themselves. The "B" received its final update in 1954 when a four speed gearbox was adopted. A steel pan seat came in 1951 but this was an option until 1954 when it also became rubber sprung.

In September 1954, the "B" became the D270. It had latterly also been available with the Perkins P3 (TA) engine. The English "B" developed 22HP on kerosine.

English "B" tractors are numbered in a separate series, the serial number being prefixed EB.

"RC" and "C"

The RC model, introduced in 1939 was basically the WC tractor with an uprated B-engine, and only 5,000 were built in the following 20-months or so. The same engine was instead taken in 1940 and fitted with a modified "B" and called the "C", which had a rowcrop front end with the capability of the choice of 3 front axle styles – wide, single front fork, or vee twin post, to take either steel or pneumatic tyres. Both the "B" and "C" were offered initially as pneumatics-only models, but with the shortages of war, both succumbed to the inevitable lugs.

English Sunset and Dusk – the "D270", "D272" and "ED40"

The "B" was restyled to become the D270 in September 1954. Although others have described the D270 as a new model, only the tinwork received a restyle and the power output was beefed up by increasing the engine speed to 1,650rpm. The D270 was now being built at the new plant at Essendine, near Stamford, Lincs. The P3/143 diesel was used as an option until 1957 when this became the P3/144 which embodied gear driven timing as on the "Fordson" Dexta.

The final variant of the "B" came in the form of the D272 in 1957 and this tractor again really only was a "B" with a new shell. Despite the common statement made that most D272s were diesels about half were in fact TVO!

The last Allis Chalmers model to appear in Britain was the ED40. This was a new model, and was just that bit heavier in build than the D272. It used a Standard-Ricardo 2.3 engine and introduced an 8-speed gearbox, 'live' hydraulics, and the option of 'live' PTO.

Depthomatic hydraulics were introduced at the Royal Show in 1963 and saw an increase in power output from the engine from 37 to 41 BHP at the same time.

The last ED40 tractors were made in 1968. Allis Chalmers never really took the UK side of the tractor market seriously, and eventually sold out their UK interests to Bamfords Ltd., who at the time were themselves controlled by the Burgess organisation, who were A-C dealers in numerous UK locations. Latter day implement production at Uttoxeter being simply a tinwork and paint disguise of the Bamford product; im most cases balers.

The "WD"

An improved version of the "WC" was introduced in 1948 and was designated the "WD". Some 131,000 units were produced between 1948 and 1953 and the tractor featured "power shift" rear wheels, a double clutch system which allowed continuous operation of the PTO and hydraulic system, and the remote control of pulled implements hydraulically could be attained. Braking was improved and a special "quick" hitch was available to facilitate the attachment of trailed implements.

In 1953 the WD45 was introduced – it was basically the same tractor as the WD but with a half inch increase in the bore. Some 83,000 of this model were produced in both gasoline and diesel forms. Allis Chalmers bought out Buda in the early '50s and this gave the opportunity to introduce direct injection 4-stroke diesels to both the wheeled and crawler ranges.

The Unusual "G"

The Allis Model "G" appeared in 1948 and was produced until 1955. Designed primarily for the market gardener or truck-farmer, it used a continental AN-62 engine with 2,375 bore and 3.5" stroke, which developed 10HP. Nearly 30,000 units were produced of this unusual looking machine.

The "CA"

The American "C" was updated and became the CA in 1950. The engine power output was pushed up by increasing the revs from 1,500 to 1,640rpm. Some 39,000 units were built up to 1957; later examples had pan seats, and the 'quick hitch' arrangements, which extended to the attachment of implements on the rockshaft type lift.

Crawlers: "The Monarch Legacy"

Allis Chalmers' entry into the crawler market started with the acquisition of the Monarch Tractor Corporation in 1928, and crawler manufacture was continued at Springfield, Illinois. The first model to bear the A-C name was the 35 which used the model "E" Allis Chalmer's built engine. In 1933 this model was renamed the "K", which was produced until 1940.

An unusual variant of the "K" appeared in 1933 in the form of the KO. This had an engine of the same bore and stroke – 5¼" by 6½" but was a development of the Waukesha Hesselman design. Whilst the engine was of low compression ratio and retained magneto/spark ignition, the fuel was delivered into the cylinders via a Bosch fuel pump and differential needle injectors.

Original "K" and "KO" tractors had steering clutches of the old "Monarch" type operated by a steering wheel but these gave way to lever operation.

The "K" series was available in two track widths giving an overall tractor width of 65.875 or 80.875 inches.

The "L" and "LO"

The biggest of the A-C crawler range introduced to replace the old "Monarch" series in 1932 was the L. This featured a six cylinder engine of 5¼" bore by 6½" stroke developing 60/80 horsepower on drawbar/belt. Initial production had used a continental engine but the six cylinder unit fitted by Allis was of their own manufacture, and could also be had in fuel oil form, making the tractor the LO. Six forward speeds were provided. An interesting feature of the "L" engine was the fitting of twin carbutettors.

The LO engine was started on gasoline and the fuel pump brought into operation once the engine had been running for a short period. The engine was also designed to idle on two cylinders to prevent carboning-up.

The L sported two exhaust stacks whilst the LO had three!

The "M"

The best known of the A-C crawlers in Britain at least was the "M" which came out in 1933 and used the UM engine common to the model "U". It was available in standard, wide track, and orchard forms and could handle 3-4 furrows. Four forward and one reverse speed were provided.

Although very much an agricultural model, the M, was used exclusively as an industrial or construction workhorse by some owners.

The "HD" Show

A new range of crawlers was introduced in 1940 and these were designed with a dual role in mind. All were equipped with the General Motors' two stroke diesel engine and three sizes were evolved, the HD7, HD10 and HD14. The HD7 had a three cylinder engine developing 54HP at the drawbar, the HD10 a 4 cylinder engine developing 80HP and the HD14 a 6 cylinder engine developing 108HP. The smallest model offered four forward speeds, and the two larger, six. The engine used a common bore and stroke of 4¼" x 5" and therefore the number of components were standardised.

In appearance also the tractors were identically styled. In publicising the new range it was expounded that you get twice the amount of work out of a two cycle diesel of the same size as the four cycle. GM engines continued to be used in the A-C crawler range until the mid '50s when Buda engines were substituted, that company having been bought out by Allis in 1952.

Forties Update: The "HD5", "HD9" and "HD15"

1948 saw the updating of the crawler range and the addition of a new model on the bottom end of the range. The HD5 embodied a GM engine developing 45 BHP from two cylinders having the same bore and stroke as previous models. It was aimed at the agricultural market, being effectively a replacement for the "M" and was therefore marketed through Totton. The big industrial crawlers in the UK were factored by Mackay Industrial Equipment of Feltham, Middlesex.

The HD5 appeared along with several model "B" tractors at the 1949 Royal Show but never took off in this country to any great extent. The General Motors engine was, incidentally, the same model GM 2-71 as adopted by Thomson for their cane conversions – a convenient choice as far as spares went.

The HD9 and HD15 models were purely updates of the previous models but when the HD19 was introduced in 1947 it had the distinction of being the first crawler tractor to be fitted with a torque convertor, a feature which was not adopted by other manufacturers for years. It was also claimed to be the world's most powerful tractor developing a maximum of 118HP at the drawbar. A 2-cycle GM six cylinder engine was used with 5" bore and 5½" stroke. In 1950 the tractor was updated and became the HD20, this incorporating an improved engine.

After Allis Chalmers acquired Buda it was politic to revamp the crawler range to use engines which were of "Allis Chalmers" manufacture and thus the use of General Motors two cycle diesels ceased.

Miscellania

Allis Chalmers did much to improve the mechanisation on smaller farms in the USA and the "All Crop" harvester was no exception in this line of activity. As easy to attach to the rear of the tractor as a reaper, available in tanker or bagger form, with or without engine, the "All Crop" heralded the start of a fashion in smaller tractor drawn combines. Allis Chalmers involvement in combine harvesters dated back to the acquisition of the Rumely line in 1931.

The "All Crop" was assembled in Britain in the '50s and caused the development of a smaller self propelled unit known as the "Gleaner". The "bagger" type was more popular in the UK and a British-built version of the B125 power unit with more compact dust guard on the radiator was used.

The unique Allis Chalmers Roto Baler was developed as a challenge to normal baling techniques. Instead of a bale chamber, the material to be baled was trapped in a series of rotating rubber belts and spun into a round bale, being ejected rearwards once a controlled size had been reached. The units were available with both PTP and engine drives, and assembly at both Totton and Essendine in the UK meant that they were to be seen at work in the hay and harvest fields of Britain.

The design was perhaps a little too advanced, yet the era of 'big balers' is now with us in the 1980s – most of these operate on very small principles to the Roto-Baler.

A major side of A-C business was that of producing and selling power units. Most of Allis Chalmers' own engines were available in this form and could usually be had either as just the engine, or with clutch, pulley, radiator and tank and suitable mounting girders. As time passed the power units became more compact from the viewpoint that the bonnet and frame were redesigned to make up one unit. The model B, WC and E engines lasted in power unit form a lot longer than the tractors themselves, not being discontinued until 1960.

1940 saw A-C move away from the use of their own engines in the larger Crawlers and the fitting of General Motors 2 stroke units. It is interesting to note that the basic GM 2-stroke diesel is little different today from those built in 1940.

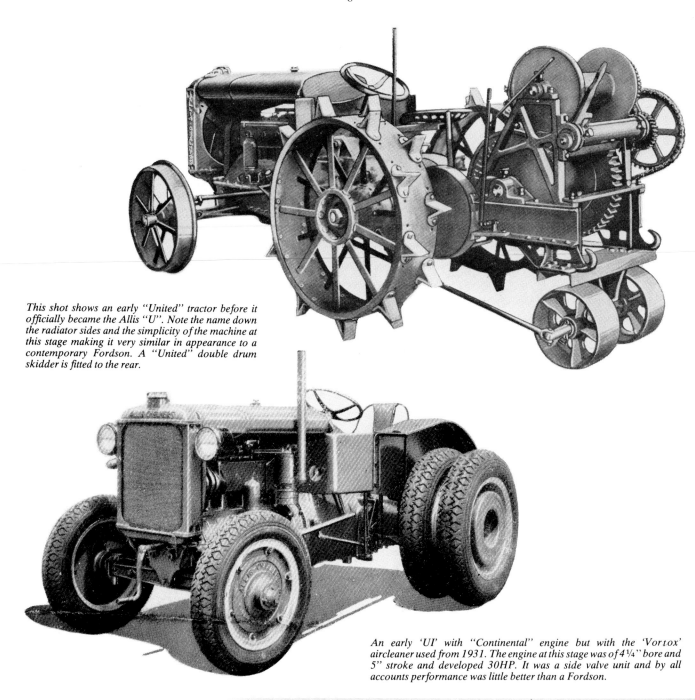

This shot shows an early "United" tractor before it officially became the Allis "U". Note the name down the radiator sides and the simplicity of the machine at this stage making it very similar in appearance to a contemporary Fordson. A "United" double drum skidder is fitted to the rear.

An early 'UI' with "Continental" engine but with the 'Vortox' aircleaner used from 1931. The engine at this stage was of 4¼" bore and 5" stroke and developed 30HP. It was a side valve unit and by all accounts performance was little better than a Fordson.

Ploughing with a 'U' Continental on the Beaver Dam farm near Laporte, Indiana.

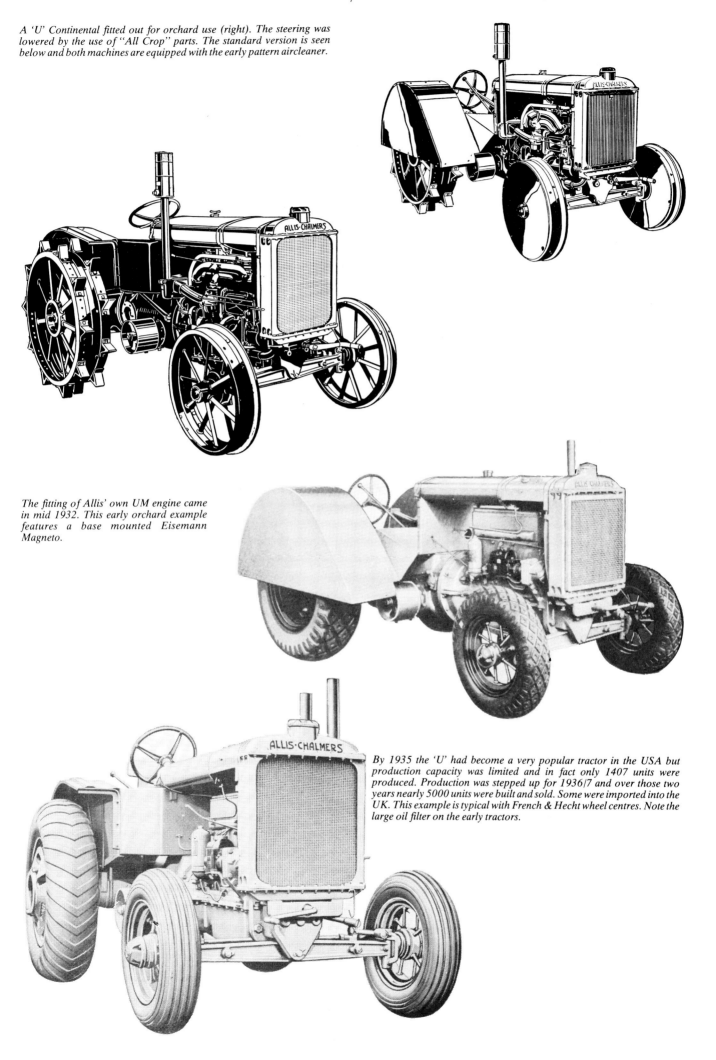

A 'U' Continental fitted out for orchard use (right). The steering was lowered by the use of "All Crop" parts. The standard version is seen below and both machines are equipped with the early pattern aircleaner.

The fitting of Allis' own UM engine came in mid 1932. This early orchard example features a base mounted Eisemann Magneto.

By 1935 the 'U' had become a very popular tractor in the USA but production capacity was limited and in fact only 1407 units were produced. Production was stepped up for 1936/7 and over those two years nearly 5000 units were built and sold. Some were imported into the UK. This example is typical with French & Hecht wheel centres. Note the large oil filter on the early tractors.

A 1935 tractor on steels is seen here and the manifold arrangements can clearly be seen. The exhaust in the forward position allowed for petrol or gasoline operation and in the centre kerosine or distillate. Also seen is the U's weak point; the band brake on the countershaft.

1935 also saw the alteration of the steering position on the tractor and the change in mudguard style as seen on this example (right). This style changed little into the war years and a wartime tractor is seen below equipped with skeleton rear wheels for test purposes by the N.I.A.E.

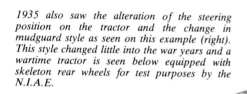

Roadless did produce some sets of DG tracks for the Allis 'U' and a tractor so equipped is seen here.

Postwar 'U' tractors were crated and assembled at Allis Chalmers' Totton plant near Southampton. Flange mounted Fairbanks Morse magneto, cartridge type oil filter, and later pattern 'can' type Donaldson Air Cleaner plus Ball Special precleaner are all featured.

6 volt lighting and starting was optional on the 'U' but most of the late imports into the UK were equipped. Late tractors also were equipped with a thermostat rendering the use of a radiator blind unnecessary.

Allis Chalmers' original "All Crop". It used the same Continental engine as the United, but the rear transmission was of a different design. To give adequate ground clearance the rear axles incorporated reduction gears at the outer ends of the cases – unlike the 'U' therefore the transmission itself was single reduction. The mechanical power lift was integral with the transmission and the operating pedal can be seen just to the right of the fender front.

With the fitting of the UM engine the 'All Crop' became the 'UC'. The slewing brakes were built into the final drive casings and were contracting bands. Operation was by hand levers which can be seen clearly in this view of a tractor with steel wheels.

The 'UC' with Air Tyres. 11.25x 24" rears and 6.00 x 16" fronts were the sizes fitted at this time.

A late 'UC' showing parallel development in details with the 'U'. The fenders were restyled in 1937.

French & Hecht supplied the wheels for the 'UC' which on later models were 28" for the rear tyres.

A 'UC' Cane model fitted with Thompsons "Hurrycane" Loading equipment.

The model 'E' started life as the 25/40 and dated from 1930. It used A-C's own engine but was handicapped by the use of a two speed transmission. This is an early example with dry aircleaner.

Later model 'E' tractors were fitted with an oilbath aircleaner mounted at the front of the engine behind the radiator on the RH side.

The 'E' was also offered on Pneumatics from 1932 – with a top speed of just over 3mph their fitment was hardly apt.

ALLIS-CHALMERS
TRACTOR DIVISION—MILWAUKEE, U.S.A.

The 'E' was replaced in 1936 by the 'A' which used a beefed up 'U' rear transmission mated to a 'K' engine with the bore reduced to 4¾. It was made for only just less than 5 years and dropped in 1941. It had a reputation of being a fuel 'guzzler'.

Early 'WC' tractors featured unlettered radiators and engines with base mounted magnetos, a large oil filter by 'Purlolator' and were available on both steel wheels (right) and pneumatics (below).

By 1935 radiator pressings were embossed, a flange mounted magneto and cartridge oil filter were adopted, and the seat was given a back rest.

ALLIS-CHALMERS
TRACTOR DIVISION—MILWAUKEE, U.S.A.

The 'WC' was A-Cs most popular seller with over 178000 units being built from 1933 to 1948. It must be remembered that the American market demanded more "Row Crop" tractors than any other type and with the added incentive of pneumatic tyres no small mid Western farm was without one! Whilst the normal configuration was vee twins a wide front axle was also available (below) and a high clearance version with single front wheel (bottom).

The 'WF' was a standard tread version of the 'WC' and lower overall weight was attained by the use of smaller final reduction housings. A variety of wheel equipment was available as shown on this page.

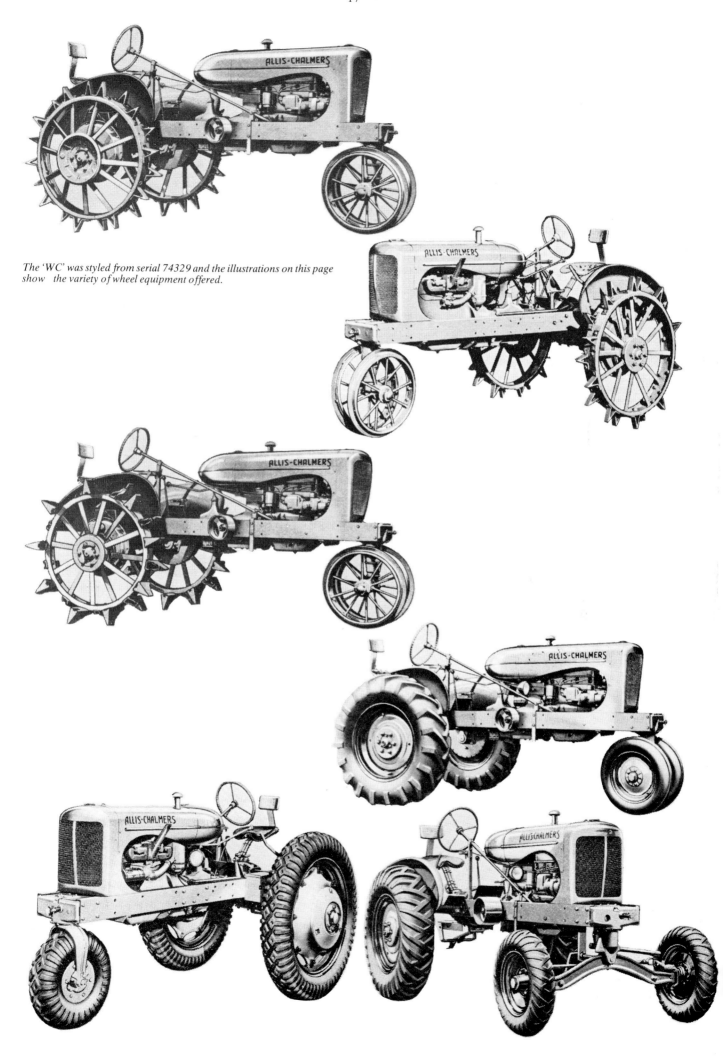

The 'WC' was styled from serial 74329 and the illustrations on this page show the variety of wheel equipment offered.

A complete range of mounted equipment was available for the WC tractor and these two shots show the mid mounted mower. Drive came from the PTO. The advantages of having a cutter bar in front of the operator were evident expressly where the mower was power driven.

The 'WF' was styled from the end of 1938 but was only produced until 1950 without any major update.

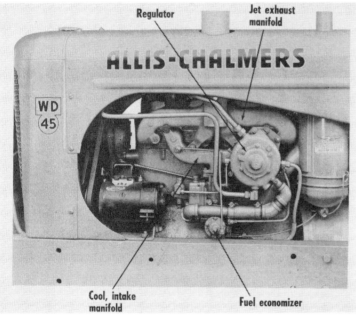

The 'WD' was an updated version of the 'WC' produced from 1948. Some examples reached Ireland – indeed one or two may have also reached the UK.

The WD45 was the final stage in development and had a bigger engine. It was introduced in 1953 and is seen here in LP Gas form. The special fuel arrangements are seen on the left. Note the power shift rear wheels.

The famous 'B' was introduced in 1937. Seen here are the "Special" (above) with lighting and starting, a basic model (right), and a mower equipped unit (below).

The 'IB' featured a subframe which carried the front axle. The two lower shots show the tractor equipped with road brush and mower.

A 'B' with single row seeder attachment driven off the rear wheel.

Above: The 'B' was really a Rowcrop machine and somewhat unsuited to the type of job it is seen performing here.
Right: 'B' with Perkins P3 (TA) engine fitted.

Below: Late English 'B' showing features exclusive to the UK build, the Lucas magneto and electrical equipment and pan type seat.

Opposite page: Top– 'B' Asparagus special with 9.00 x 36" rears. Centre– Potato special with 28" rears. Bottom: English built 'B' with A-C's own three point linkage and drawbar. Note the steel pan seat.

Above: The 'C' was simply a 'B' with a modified front end which could take three axle styles. Offered initially with pneumatics only, wartime shortages created a steel wheeled version as can be seen above being loaded onto railway flat cars at the Milwaukee plant.
Below: Row crop work was the essential selling point of the 'C' and this example is seen fully equipped with a two row seeder unit.

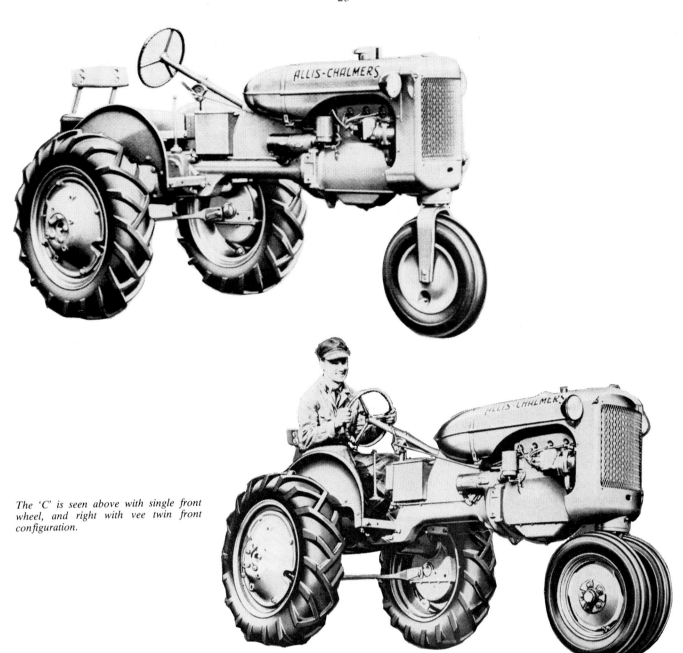

The 'C' is seen above with single front wheel, and right with vee twin front configuration.

The 'RC' preceded the 'C' by just over a year and used the 'WC' frame with a bored out 'B' engine. It accounted for only 5000 units before the uprated engine was used in the 'C'

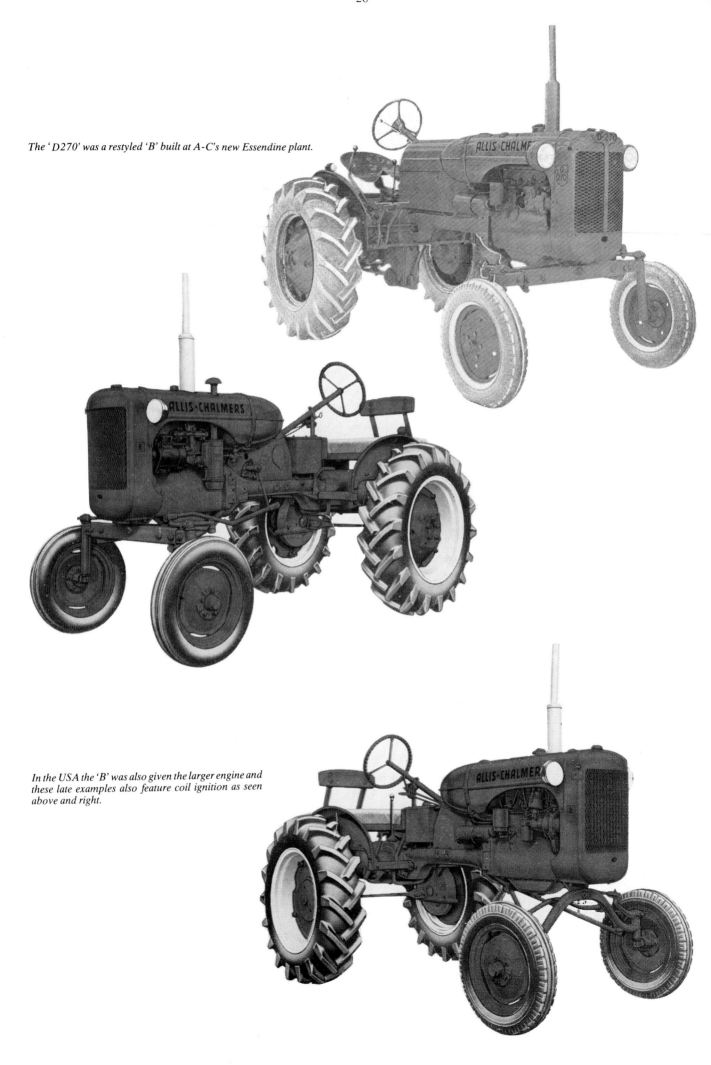

The 'D270' was a restyled 'B' built at A-C's new Essendine plant.

In the USA the 'B' was also given the larger engine and these late examples also feature coil ignition as seen above and right.

The 'D272' was the final British essay on the 'B' and was available with the Perkins P3/144 engine as seen (right).

The 'D272' was also available in high clearance form (left). Whilst a proportion of 'D272' tractors had diesel engines nearly 50% were fitted with A-C's own TVO units (below).

An unusual addition to the A-C range was the '6' which had a rear mounted engine by Continental and a full range of implements were available to fit the "tool bar".

The 'ED40' was the last British A-C model and was engined by the Standard Motor Co., who had a surplus of diesel engines to dispose of when Ferguson went "Perkins" in 1958. This is a late example with a Winsam Cab and front end loader.

The 'ED40' was simply an overgrown 'B' with the unfortunate addition of the Standard Ricardo engine which had been such a disaster in the Ferguson FE35.
Baling in Lincolnshire in the mid '60s.

An 'M' crawler in its standard form is seen here with the 1935 version below and the 1938 version to the right.

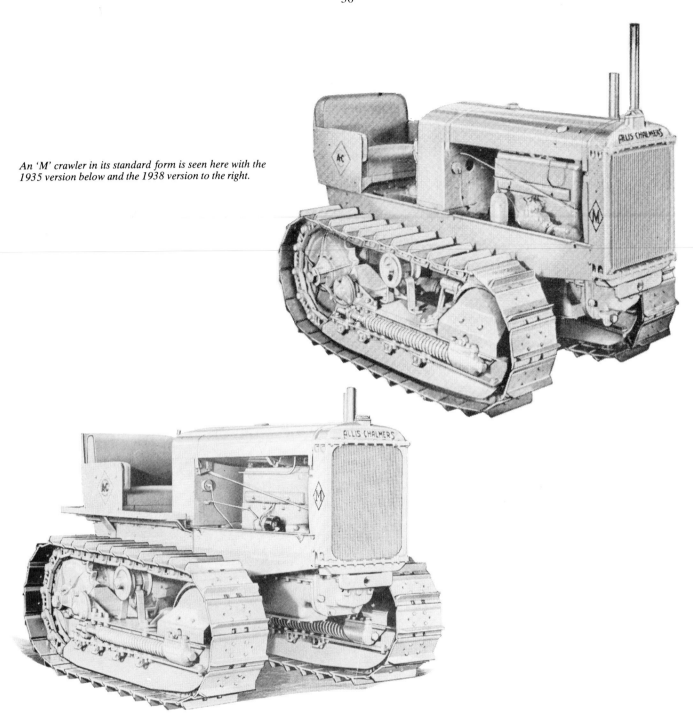

An orchard model was also available and is seen here complete with track guards and low rear driving position.

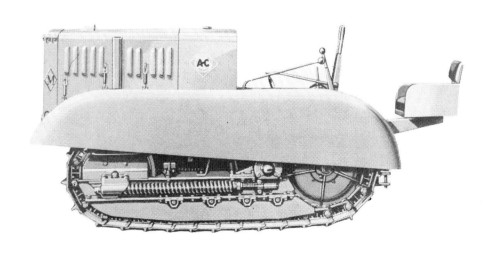

The '35' crawler used A-C's own engine as was a direct descendant from the 'Monarch' range.

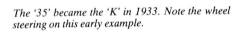

The '35' became the 'K' in 1933. Note the wheel steering on this early example.

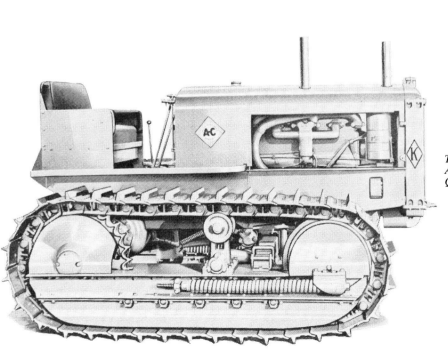

The 'KO' was designed to run on diesel fuels using A-C's own development of the Qaukesha-Hesselman engine.

A-C's big crawler model was the 'L' and very impressive it looked too. Initial production used Continental engines but A-C's own six cylinder unit was soon substituted.

The 'LO' used A-C's own fuel oil engine of Waukesha-Hesselman design.

Rear view of the 'L'.

The 'S' and 'SO' crawlers were A-C's last essay in using their own engines until the mid 1950s. in crawler models, and the example shown here is fitted with a Baker Bulldozer.

Baker Bulldozers also made the 'K' (right) and 'L' (below) models into units for the construction industry.

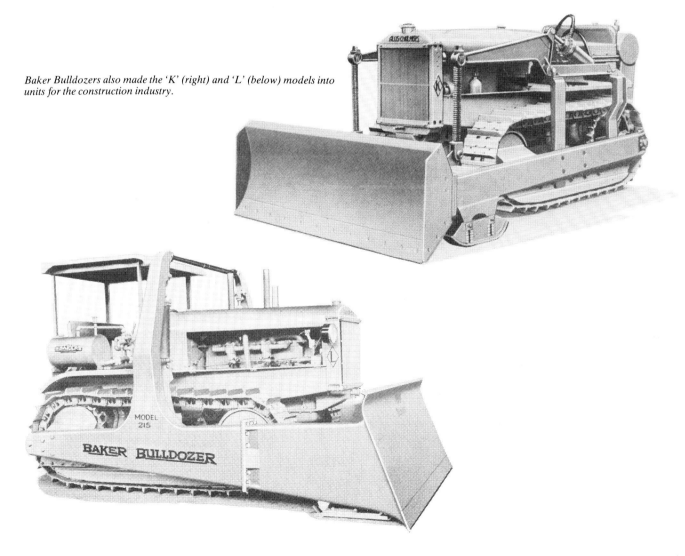

The HD 7 had a GM three cylinder engine.

The HD 10 had a GM four cylinder engine.

The HD 14 had a GM six cylinder engine. All three models shared common components and a family resemblence.

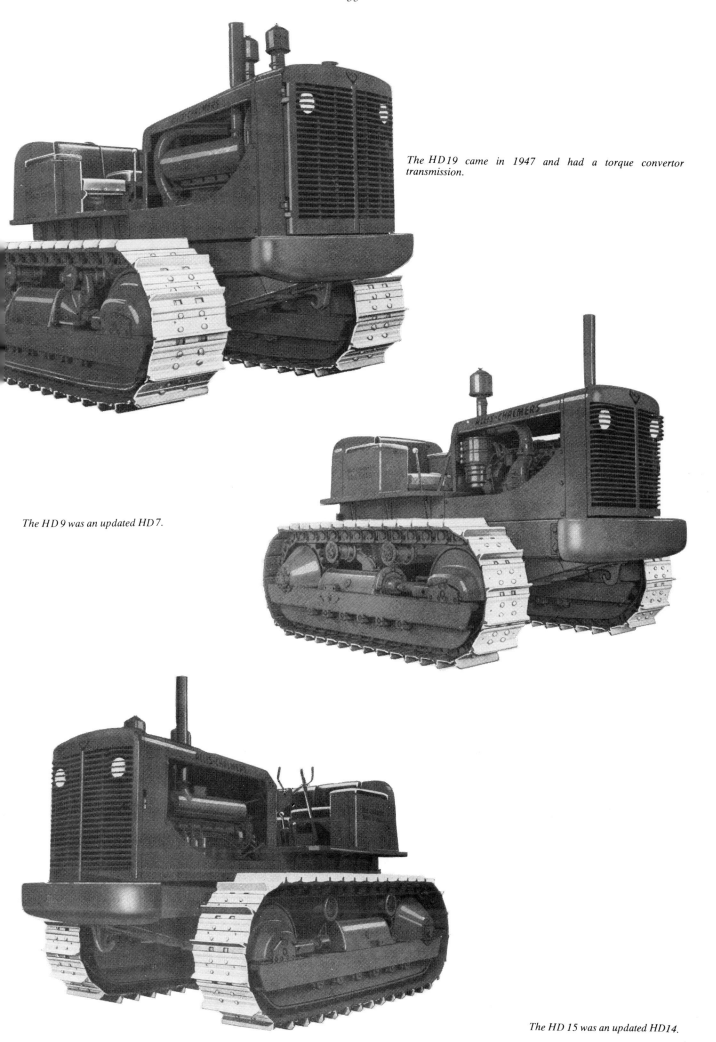

The HD19 came in 1947 and had a torque convertor transmission.

The HD 9 was an updated HD 7.

The HD 15 was an updated HD14.

The HD5 as introduced in 1948 and had a GM two cylinder engine. It was intended to fill the gap left by the cessation of 'M' production and was sold in the UK as an agricultural unit.

This shot shows the N.I.A.E. workshops in the late 1940s with two classic American tractors well known in the UK. Note that both examples have pneumatic tyres with UK manufactured centres.

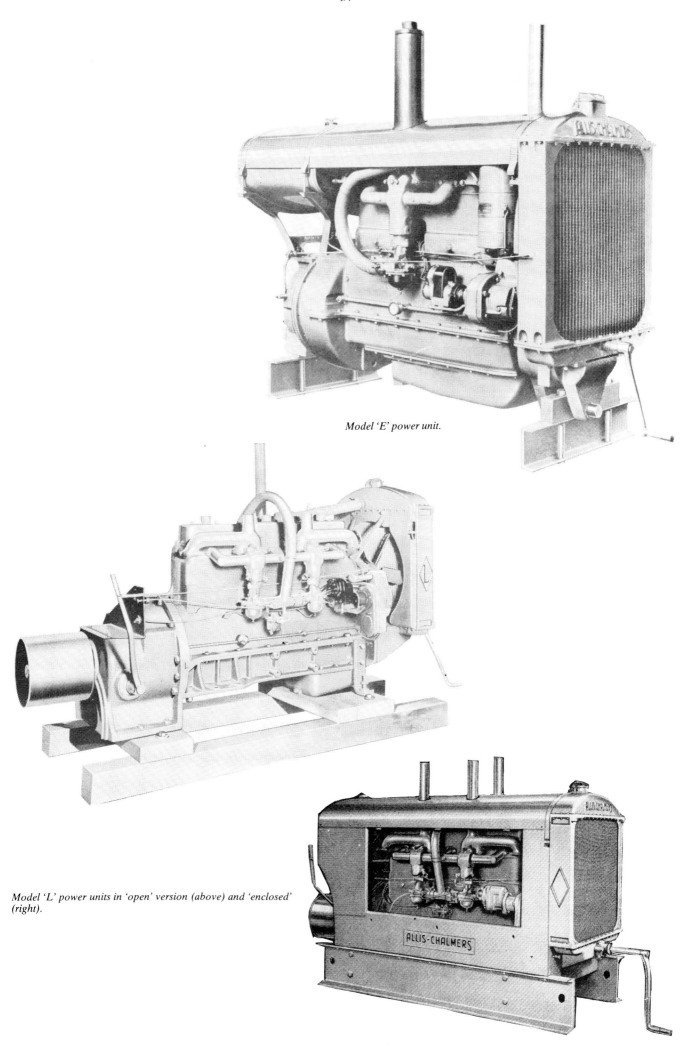

Model 'E' power unit.

Model 'L' power units in 'open' version (above) and 'enclosed' (right).

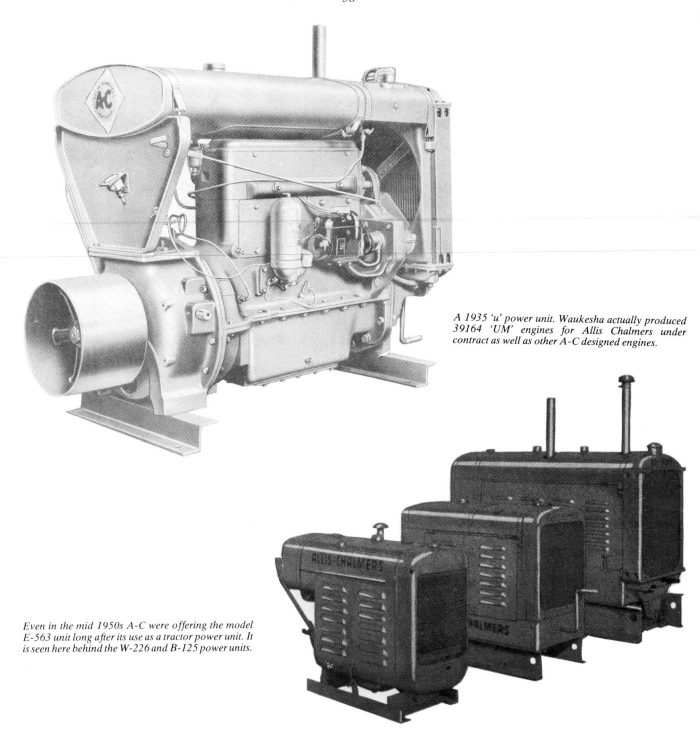

A 1935 'u' power unit. Waukesha actually produced 39164 'UM' engines for Allis Chalmers under contract as well as other A-C designed engines.

Even in the mid 1950s A-C were offering the model E-563 unit long after its use as a tractor power unit. It is seen here behind the W-226 and B-125 power units.

The All Crop 60 Combine as a bagger for the UK market with B-125 power unit.

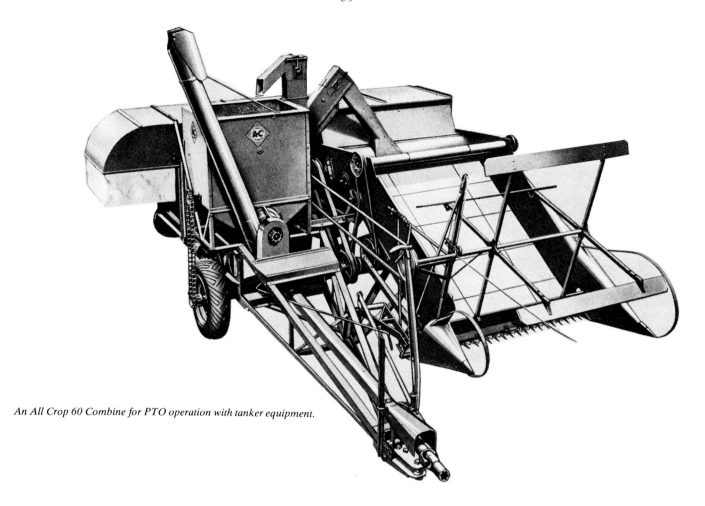

An All Crop 60 Combine for PTO operation with tanker equipment.

A PTO driven Rotobaler.

Allis Chalmers Serial Numbers

United States Production West Allis Works. (wheeled tractors)

Model E 25-40.
1930 .... 24186 to 24842
1931 .... 24843 to 24971
1932 .... 24972 to 25023
1933 .... 25024 to 25061
1934 .... 25062 to 25308
1935 .... 25309 to 25581
1936 .... 25582 to 25611
None since
Tractor serial no. is on transmission housing in front of gear lever.

Model U Continental.
1929 .... U-1 to U-1974
1930 .... U-1975 to U-6553
1931 .... U-6554 to U-7261
1932 .... U-7262 to U-7404
None since with Continental Engines.

Model U (A-C UM Engine).
1932 .... U-7405 to U-7418
1933 .... U-7419 to U-7684
1934 .... U-7685 to U-8062
1935 .... U-8063 to U-9470
1936 .... U-9471 to ....
1937 .... to U-14854
1938 .... U-14855 to U-15546
1939 .... U-15587 to U-16077
1940 .... U-16078 to U-16721
1941 .... U-16722 to U-17136
1942 .... U-17137 to U-17469
1943 .... U-17470 to U-17801
1944 .... U-17802 to U-17819
1945 .... 18022 to 18234
1946 .... 18235 to 20773
1947 .... 20774 to 21021
1948 .... 21022 to 22127
1949 .... 22128 to 23028
1950 .... 23029 to ....
1951 .... 22548 to ....
1951 .... 22561 to 22823
1952 .... 20586 to 20748
1952 .... 22824 to 23056
Towards the end of production there are inaccuracies in the recording of serial numbers. Some tractors may have been built in 1953 to special order.

Model UC/ 19-30 Continental
1930 .... UC-1 to UC-38
1931 .... UC-39 to ... UC-1099
1932 .... UC-1100 to ... UC-1231
1933 .... UC-1232 to ... UC-1268
None since with Continental engine.

Model UC (A-C UM Engine).
1933 .... UC-1269 to ... UC-1293
1934 .... UC-1294 to ... UC-1551
1935 .... UC-1552 to ... UC-2000
1936 .... UC-2001 to ... UC-2770
1937 .... UC-2771 to ... UC-3756
1938 .... UC-3757 to ... UC-4546
1939 .... UC-4547 to ... UC-4769
1940 .... UC-4770 to ... UC-4971
1941 .... UC-4792 to ... UC-5037
1951 .... 5806 to 5938
Although UC production ceased in 1941, these tractors were built to special order until 1951, usually for Cane conversions. After this date, Thompsons used UC transmissions and added GM diesel engines until they adopted Fordson units in the mid-fifties.
Location of serial number. This is to be found on the rear of the differential housing on all U, UC and UI tractors.

Model WC
1933 .... WC-1 to WC-28
1934 .... WC-29 to ... WC-3126
1935 .... WC-3127 to . WC-13869
1936 .... WC-13870 to . WC-31783
1937 .... WC-31784 to . WC-60789
1938 .... WC-60790 to . WC-75215
1939 .... WC-75216 to . WC-91533
1940 .... WC-91534 to 103516
1941 .... 103517 to 114533
1942 .... 114534 to 123170
1943 .... 123171 to 127641
1944 .... 127642 to 134623
1945 .... 134624 to 148090
1946 .... 148091 to 152844
1947 .... 152845 to 170173
1948 .... 170174 to 178202
Model replaced by WD.

Model WF
1937 .... WF-1 to WF-288
1938 .... WF-289 to ... WF-1335
1939 .... WF-1336 to ... WF-1891
1940 .... WF-1892 to ... WF-2299
1941 .... WF-2300 to ... WF-2703
1942 .... WF-2704 to ... WF-3003
1943 .... None built.
1944 .... WF-3004 to ... WF-3194
1945 .... 3195 to 3509
1946 .... 3510 to 3747
1947 .... 3748 to 4110
1948 .... 4111 to 5499
1949 .... 5500 to 7317
1950 .... 7318 to 8315
1951 8316
None since.
Serial number of WC and WF tractors is on rear axle housing. Electrics were standard from WC-74329 and WF-1904.

Model WD
1948 .... 7 to 9249
1949 .... 9250 to 35444
1950 .... 35445 to 72327
1951 .... 72328 to 105181
1952 .... 105182 to 126981
1953 .... 127007 to 131242
1954 .... 131243 to 160384
None since.
Tractor serial number is on top of left hand brake housing.

Model WD45
1953 .... 146607 to 160385
1954 .... 160386 to 190992
1955 .... 190993 to 217991
1956 .... 217992 to 230294
1957 .... 230295 to 236958
Model WD and WD45 tractors were built concurrently during 1953 and shared the same number series.

Model A
1936 .... 25701 to 25725
1937 .... 25726 to 26304
1938 .... 26305 to 26613
1939 .... 26614 to 26781
1940 .... 26782 to 26895
1941 .... 26986 to 26914
1942 .... 26915 to 26925
None since.
Serial number is on differential case.

Model B (US production) Continental Engine.
1937 .... 1 to 96

Model B (US production) Allis Engine.
Serial prefix B
1938 .... 97 to 11799
1939 .... 11800 to 33501
1940 .... 33502 to 49720
1941 .... 49721 to 56781
1942 .... 56782 to 61400
1943 .... 64501 to 65501
1944 .... 65502 to 70209
1945 .... 70210 to 72264
1946 .... 72265 to 73369
1947 .... 73370 to 74079
1947 .... 75080 to 80555
1948 .... 80556 to 85833
1948 .... 87834 to 92294
1949 .... 92295 to 102392
1950 .... 102393 to 103578
1950 .... 106579 to 114526
1951 .... 114527 to 118673
1952 .... 118674 to 122309
1953 .... 122310 to 124201
1954 .... 124202 to 124710
1955 .... 124711 to 126496
1956 .... 126497 to 127185
1957 .... 127186 to ....
None since.
Model B tractors were not imported to the UK after 1942.

Model EB.
Serial prefix EB.
Accurate information is not available for this model or any British Toton or Essendine assembled tractors.